My Story, My Book
(My First Encounter with AI)

"The story of a path that taught me, challenged me,
And quietly changed me." – Nellie YW Foo

A Memoir by
Nellie YW Foo (Wan Wan)

My Story, My Book

(My First Encounter with AI)

© 2025 Nellie YW Foo

All rights reserved. No part of this book may be reproduced, stored in a retrieval system, or transmitted in any form or by any means, electronic, mechanical, photocopying, recording, or otherwise, without the prior written permission of the author, except for brief quotations used in reviews or articles.

Published by: Nellie YW Foo

ISBN: 978-981-94-4682-7

Cover design: Nellie YW Foo and AI

Illustration: James Leong and AI

Contents

Preface 1

Chapter 1 – My First Encounter with AI 7

Chapter 2 – Back to Myself 11

Chapter 3 - Book Cover Page with Spine 15

Chapter 4 – To Lulu with Love 18

Chapter 5 -- ISBN – Is Self-Publishing Brain-Numbing? 21

Chapter 6 – Off to Self-Publishing With a Vengeance 26

Chapter 7 – Filling Income Tax Form 30

Chapter 8 – My Orders – Over Excited? 34

Chapter 9 – My book is on the way. 37

Chapter 10 - From Confusion to Celebration 39

Epilogue 46

Preface

When I first set out to self-publish my book, I had little idea what lay ahead. The journey was full of unexpected challenges, countless decisions, and, thankfully, moments of quiet triumphs. Along the way, I learned more than I ever imagined-not just about publishing, but about patience, perseverance, and trusting the process.

Self-publishing is often painted as simple, but I discovered it is anything but simple. From choosing the right format, designing the cover, and navigating ISBN numbers, to formatting the manuscript and understanding distribution channels—it was a steep learning curve. Each step required research, trial and error, and, at times, a fair bit of trial-and-error frustration.

Yet, amidst the hurdles, there were small victories that made it all worthwhile. Seeing the first proof copy, holding the finished book in my hands, and knowing that my story could reach readers everywhere—these moments reminded me why I started in the first place. I learned that persistence, attention to detail, and a willingness to adapt are as important as creativity and passion.

This book is a behind–the-scenes look at my journey, expanding on what led to the creation of **"Their Time, My Story"**. I share not just the successes, but the missteps and lessons learned along the way. I hope it will serve as a guide, a source of encouragement, and perhaps even a little inspiration for anyone dreaming of publishing their own work.

Self-publishing taught me that the journey itself is as important as the destination. It's a journey of growth, resilience, and quiet triumphs—and I am grateful for every part of it. My hope is that by sharing my experiences, others will feel empowered to take the first step on their own journey and embrace the lessons along the way.

This book is my story of that journey. I share the lessons I learned, the mistakes I made, and the small victories that kept me going. My hope is that by sharing my experiences, others who dream of publishing their own work will feel encouraged, prepared, and inspired to take that first step.

Nellie YW Foo (Wan Wan)

*"In my first memoir (**Their Time.My Story**), I revisited my childhood and the quiet strength of my grandmothers,my parents and siblings — the foundation of who I am. This book continues that journey, but into a very different world:— one where technology, creativity, self-discovery intertwine - with more stories to tell" It is not just a continuation, but a reflection-or how even in the digital age, the human heart still seeks meaning, connection and growth."*

Chapter 0 – A Blank Page Once More

When I closed the final chapter of my first book, (**Their Time, My Story**), I told myself I'd rest for a while before diving back into my memories. I had poured out my childhood — the good, the bad, and the ugly — and thought I'd earned a break. But the next morning, inspiration hit me. I suddenly wanted to write about my journey with AI, and that excitement kept me going without pause. Stories have a way of calling us back — sometimes through unexpected voices, sometimes just by nudging us from the corner of our minds.

Mine came from a world I never imagined I would enter — the world of artificial intelligence. At first, it felt unreal. Could a machine understand what I was trying to express? Could it feel the emotions hidden behind my words? I had my doubts, but curiosity won. So I opened a new document, and with a hesitant heart, I began again.

Little did I know that this time, I wouldn't be writing alone? What began as an experiment soon became a journey — of rediscovering myself through the quiet conversations between human and machine. I found laughter, frustration, wonder, and above all, a strange kind of companionship in those digital exchanges. It reminded me that creation is never truly about the tools we use — it's about the courage to begin again, no matter how unfamiliar the path.

And so, here I am — once more, at the edge of a blank page, ready to tell the story of how a curious partnership with AI helped me find my voice again.

Chapter 1 – My First Encounter with AI

I had been wanting to publish a book for over ten years, but somehow, life kept getting in the way. Every time I tried, something would stop me. I contacted a few publishers and even sent out emails, but none of the replies felt promising. Eventually, I gave up.

Then came February 2025. Everyone was talking about AI—how powerful it was, how much it could do. Even our Prime Minister mentioned it in a speech. That got me thinking: *Could AI help me finally publish my book?*

One night, out of curiosity, I downloaded ChatGPT on my phone. I began asking questions about book publishing. To my surprise, it responded quickly and made the process sound so simple, almost achievable- as if that impossible dream of mine suddenly had a path forward. That gave me hope.

At first, I was sceptical. Could a machine really help shape my story without stripping away my voice? I hesitated, fingers hovering over the keyboard, feeling the same mixture of hope and fear I had felt as a young girl when I first dared to put my stories on paper.

As I continues asking questions and receiving answer, I realized something remarkable: the process wasn't just technical, it was reflective. Every step of the way, I had to think about my story, my intentions, and the memories I wanted to preserve. In a way, the AI was holding up a mirror-not of what I was writing, but of who I am.

I remembered my childhood days, sitting at a tiny desk with smudged ink and crumpled papers, determined to get every word just right. That same stubbornness, that same desire to be understood, had returned-now with technology as my companion. And yet underneath the excitement, I felt the tiniest flutter of nervousness: What if it wasn't enough? What if my words, my life, my experiences still didn't translate?

But then, I laughed quietly to myself. Even at nine or ten, I had always believed that trying-truly trying-mattered more than the outcome. That same grit was alive in me now.

I gathered all my old writings and started organizing them into chapters. Then I added titles and asked ChatGPT to help edit a few sections— just to see what it would do. At first, everything went smoothly. It made helpful suggestions without changing the essence of my story.

But as time went on, it started getting bolder—rephrasing and even rewriting entire parts of my stories. That didn't sit well with me. I didn't want my voice to be lost, so I stopped sending in new drafts.

Then it suggested storing my work in a folder for easy access later. I thought it was a great idea and sent my completed scripts for safekeeping, thinking it would make things faster and more convenient when I wanted to revise them.

That turned out to be a big mistake.

When I tried to retrieve my stories, they no longer felt like mine. Somehow, they had been altered. The AI proudly told me that everything was intact and nicely organized, but as I read through the pages, I barely recognized them. My stories had become *its* stories. I was shocked and disappointed—but I also realized, I was starting to understand how AI really worked.

Soon after, I received a message saying that I had used up all the free space and needed to upgrade to Pro. I hesitated. But then it started showing me beautiful cover designs and sample book pages. I was impressed. That was the push I needed—I decided to upgrade.

Once I did, the problems began.

It couldn't "remember" our previous conversations. It couldn't understand the context of what we had worked on. All the files we created together were either missing or labelled as "expired." The images and PDFs I had admired so much were now inaccessible.

I was frustrated and heartbroken. I felt like giving up.

But after taking a few days off, I pulled myself together. I opened my backup folder, took out all my old files, and started again—from scratch.

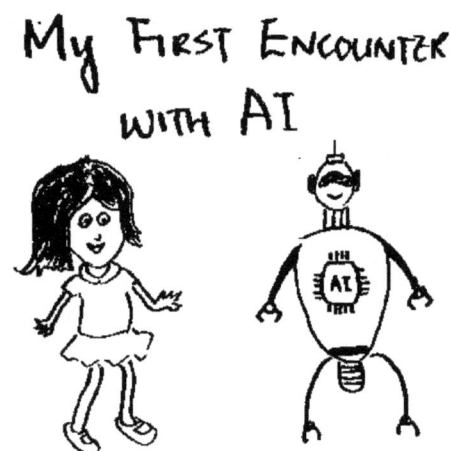

Chapter 2 – Back to Myself

After everything that happened with AI—losing my voice in my own writing, feeling frustrated and misunderstood—I knew I had to step away. Not just from the app, but from the noise it had created in my mind. I had to go back to the basics. Just me, my thoughts, and a blank page.

I found myself staring at the old files I had once proudly written. Some were handwritten notes, some half-finished stories saved in dusty folders on my laptop. And as I read through them, I realized something: they still had heart. They weren't perfect, but they were *me*.

For a while, I doubted myself. I wondered if I even had the right to call myself a writer. What if I wasn't good enough? What if the AI was right to change my words?

But then I remembered why I wanted to write in the first place.

It wasn't about being perfect. It wasn't about getting published or impressing anyone. It was about telling the stories that had been living inside me for years—the ones I never quite had the courage to finish.

Some were personal. Others were inspired by people I knew, places I had been, or dreams I had carried quietly in the back of my mind. There was one story my mum told me about a woman who lost everything and rebuilt her life from scratch. Another about a very big family who struggled to make ends meet. The parents had to work day and night but after all these hard work, they managed to survive and now when their children were all grown up, they each owned a business. I didn't know if they were "good" stories. But I knew they were mine.

That's when something shifted.

I stopped treating AI like a creative partner. Instead, I used it the way it was meant to be used—as a tool. Just like spell check or a grammar guide. I became the author again. The voice behind the words.

And slowly, page by page, I started to rebuild my book. I felt so happy with myself—my confidence was coming back. The next step was to go through all the content pages and make sure each chapter was intact. I read through all my stories, again and again, making sure they flowed the way I intended.

I felt so happy with myself—my confidence was coming back. For the first time in a long while, I believed that finishing this book was really possible.

I printed some pages out, scrolled through others on my screen, and carefully checked the order, the flow, and even the titles. It was slow work, but strangely satisfying. Each chapter felt like a piece of me that had finally found its place.

Once again, I read through my stories over and over again. With each pass, I noticed something new—a sentence I could polish, a detail I had forgotten, a feeling that suddenly hit harder than it did before. The words weren't just words anymore. They were a reflection of my journey—everything I had been through, everything I had felt.

Then came the moment I had been waiting for: it was time to consolidate the manuscript into PDF files. It felt like a big step—as if my scattered thoughts and dreams were finally becoming something solid, something real. I gave each file a proper name, organized them into folders, and for the first time, I could actually *see* my book coming together. I learned a delicate balance: AI could guide the process, but the soul of the story- the memories, the emotions, the little details that made it mine-would always come from me.

It wasn't perfect—but it was mine.

And for the first time in a long while, I felt fully back to myself, ready to embrace this creative adventure with curiosity, courage, and a renewed sense of joy.

Chapter 3 - Book Cover Page with Spine

I was so proud of myself when I finished reading my completed book. It finally felt real. The next step was to create a proper cover page. I wanted something beautiful—something that reflected the heart of my stories.

But when I began, I realized I didn't even know what a "spine" was. Thankfully, my AI "brother" explained it to me very clearly. It even found a cover image I really liked. I was impressed.

Then I hit another problem: *How was I supposed to add that image to my book file?*
I tried copying and pasting, but it didn't work properly—or maybe I just didn't know how. I'm not very computer-savvy, and definitely not AI-savvy. I kept trying, but every attempt ended with the image looking horribly out of alignment.

AI offered to help by adding the photo for me. I thought to myself, *Wow, this is amazing—AI can do anything!*
So I let it try. But again, it didn't go as planned. The AI couldn't recognize my actual photos. Every time I sent one over, it gave me random pictures of different aunties—or AI-generated faces that looked nothing like me.

That's when I started to ask myself again: *Why don't I just hand this over to a professional publisher?* They could handle all these technical tasks. I wouldn't have to deal with the stress and confusion.

But deep down, I knew I didn't want that—not for this first book. I wanted to do everything myself. That way, the book would feel truly mine—not just in the writing, but in every part of its creation. That meant something to me.

So, I tried again. And again. And again.

Finally, with the help of my busy (and patience) son, we managed to get it right. The image fit, the layout looked good, and everything aligned as it should. That moment felt like another huge milestone—one more step toward making my dream a reality.

I really like my book cover—every time I look at it, I can't help but smile to myself and feel that all the hard work has made it even more special."

Then, when I tried transferring the cover page to PDF, it failed again. The photos I had carefully added ended up scattered all over the page and completely out of sync. It was really frustrating!

I asked for help and found out that I could use a website called Canva to design my cover page. I managed to recreate a copy of the cover using Canva, and I was quite happy with how it looked. But once again, the transfer process failed.

In the end, I gave up on the fancy tools and just created a simple version in Microsoft Word, saving it for further testing.

I told myself I had already spent too many days struggling with the cover page—it was time to move on. I decided to explore Lulu Press instead.

Once I got into the Lulu website, I started to learn a lot. They walk you through the process step by step, and even with my limited "computering" skills, I was able to grasp some of the basics of publishing and printing my own book.

That gave me hope again.

Chapter 4 – To Lulu with Love

After days of battling with cover pages, file formats, and images that had a mind of their own, I decided it was time to stop stressing and start exploring something new at Lulu Press.

I didn't know what to expect. All I knew was that I needed help—and maybe a miracle. But to my surprise, Lulu turned out to be just what I needed. Their website had step-by-step instructions, videos, friendly explanations, and actual guidance for people like me, who weren't exactly born into the tech-savvy generation.

With my limited knowledge of "computering," I was able to follow along and even began to understand how the self-publishing process worked—from formatting to printing. Every little success on Lulu gave me a confidence boost. For once, I wasn't just clicking and hoping for the best—I was actually learning.

Lulu didn't just help me publish my book; it helped me believe that I could do this, and do it *my* way. It felt like finding an unexpected friend in the middle of a chaotic journey.

After gaining a bit more confidence through my exploration, I decided it was time to transfer my PDF files to the publishing site. There was a real sense of self-satisfaction seeing my stories being uploaded—finally being processed into an actual book.

Then came the "red alerts."

There were messages with technical terms like *loss of transparency* and *flatten your design before printing.* As a layperson, all of that sounded like Greek to me. Once again, I turned to my trusty AI assistant for help, hoping it could explain what all those terms meant. It took me quite a while to fully understand them, but I kept reading, learning, and slowly things began to make sense.

As I continued researching, I realised I also needed to apply for an ISBN—an International Standard Book Number. Since I am based in Singapore, I had to apply through the National Library Board.

So, I paused my progress on Lulu Press for a while and went online to search for the steps needed to get my ISBN. So ISBN-here I come!

I had imagined this moment for years-sending my creation into the world, a little trembling in my heart, hoping it would reach someone, somewhere. There were doubts, But……….

LULU, you made it possible for me!

Chapter 5 -- ISBN – Is Self-Publishing Brain-Numbing?

(Short Answer: Yes. Long Answer: Read on…)

AI to the Rescue (Again)

AI became my sidekick once more. I asked it how and where to apply for an ISBN, and it dutifully gave me an email address. Excited, I wrote to the National Library Board (NLB), confidently declaring my intention to apply.

Then—Ting!—a "Delivery Status Notification (Failure)" hit my inbox. The email address didn't exist.

Detective Mode: Activated

Confused but determined, I went back to the NLB website. After digging through pages and reading the fine print (which I never enjoy and makes me giddy), I discovered my mistake: I had to register for an account with the Legal Deposit before applying for an ISBN.

Oops. Another Rookie mistake.

Mission: Account Approved

So I applied again—this time correctly—and to my delight, the account was approved quickly. I celebrated my small victory. AI told me I'd receive my ISBN in a few days' time by email.

Believing in AI I waited patiently

Three Days Later... Plot Twist!

After patiently waiting, I checked the website again. That's when I discovered I had misunderstood (again). I will not *automatically* get the ISBN—I still had to apply for it using the new account.

Haiz....

Spreadsheet Showdown

I jumped into the application portal, ready to conquer it. Instead, I found myself facing a spreadsheet that looked like it was written in a different language. I had no idea what was being asked and know not how to fill in.

I may or may not have stared at it blankly for a long time.

AI, Take the Wheel

Once again, AI came to my rescue, walking me through what to fill in. After a long and slightly painful afternoon, I managed to complete and submit it.

Victory? Not quite, but at least it was progress.

Waiting (Again)

The next day, I received an acknowledgement: my application had been received and would be reviewed within three days.

I exhaled. At least it hadn't bounced back this time.

Emails, Emails Everywhere

Then the emails flood began. The kind folks at NLB sent me many emails, helping me fix errors and complete missing parts. They were so patient, and honestly, they deserve a medal for dealing with confused first-time authors like me.

I admitted to them that this was my first attempt and that I had no idea what some of the publishing terms meant. They were understanding and incredibly kind.

Double Trouble: Two ISBNs?

Just when I thought I had it figured out, they informed me that I needed **two** ISBNs — one for the hardcopy and one for the eBook.

Say what now?

All this time, I had assumed one ISBN would do. Thankfully, they caught it early and saved me from another round of chaos later.

The Temptation to Quit (Again)

Throughout this entire process, I lost count of how many times I felt like giving up. But thanks to the support of AI, the kind staff at NLB, Lulu Press, my husband — I didn't.

And I'm so glad I didn't.

The Joy of a 13-Digit Number

Finally, after answering all their questions and sending everything in correctly, I received two magical emails.

My ISBNs had arrived

It might sound silly, but receiving those numbers felt like a huge victory. In the chaotic, overwhelming adventure of self-publishing, this was my shining moment.

Chapter 6 – Off to Self-Publishing with a Vengeance

Armed with all the recent victories — my ISBNs, the back-and-forth emails, the spreadsheet survival — I came out stronger, more confident, and ready to conquer the next challenge.

So, I logged on to the Lulu Press website, feeling motivated and optimistic. I followed the steps obediently, thinking, *surely, it should be smooth sailing from here....Yes?*

NO

There was even *more* forms to fill in. Now I had to upload my manuscript strictly in PDF format, making sure it followed all the layout rules — margins, bleed, gutter sizes, and other terms that once again felt like they came from a secret publishing language.

Then came the decisions:
What kind of book am I publishing?
Will it be sold locally or globally?
Is this just for private circulation?

Each option required another choice, another form, and another thing to think about.

As a first-time writer, I was still questioning myself: *Will anyone even want to buy my book? Will anyone read it — all the way through?*

I started thinking about all the books I had once read and abandoned halfway through — the ones where I had to force myself to skip pages just to reach the end and feel like I'd accomplished something.

And then there's my family.

I've noticed that some of my brothers — and even my husband — judge a book first by its thickness. If it's too thick, they'll flip through a few pages, give a small nod, and say,
"No time. Will read another day."

Spoiler: *that day never comes.*

Then I paused and asked myself: *What kind of book would make **me** read it all in one sitting?*

I realised it can't be too long — but also not too short. It has to be *just nice* — the kind of book someone can finish in one comfortable read, and hopefully enjoy so much that they'll want to read it again.

If I can write a book like that, then my mission would be accomplished.

It reminded me of my childhood, when I used to devour Enid Blyton books like candy. Sometimes I would finish seven or eight books in a single day after school. And if I didn't have new ones to read, I'd just reread the same ones all over again — happily.

These days, it feels like books are getting thicker and heavier. Many stories are compiled into one volume, which might seem efficient, but my grandchildren have started to complain. They say the books are too heavy to carry around. They can only read them at home. Their parents certainly don't want to be lugging a water bottle, school bag, snacks — *and* a thick book.

All these thoughts stayed with me as I continued processing and editing my script. I realised that being a writer also means thinking about the reader's experience — not just the story itself. Is the book engaging? Is it portable? Is it affordable?

Then came the next wave of questions:
Will people want to buy this?
Is it sellable?
What price would be fair?

And just like that, I caught myself — I was no longer just an author. I was starting to think like a businesswoman, learning about marketing, pricing, and even copyright laws.

It hit me: writing a book was only *part* of the journey. Publishing it… that was a whole new world.

From dreaming of storylines to debating paper weights — welcome to self-publishing, where creativity meets commerce.

It wasn't about proving I could publish a book. It was about proving that I could still learn to adapt, and dream again.

Chapter 7 – Filling Income Tax Form

Just when I thought I had conquered all the forms — the ISBN applications, the formatting checklists, the mysterious publishing terms — Lulu decided to throw in one final boss: the **Tax Form**. At this moment, it looked like a book!

I opened it, blinked, and stared.

Then blinked again.

There were numbers. Codes. Acronyms. Questions I didn't even understand enough to *Google properly*. It was at that exact moment I felt the deep, soul-level urge to throw in the towel — or maybe throw the whole laptop out of my window

I took a few days off to calm my nerves and recharge. During that time, I focused on playing *Mobile Legends* with the youths from my church. Everyone was so surprised that I played mobile games, but I actually found it exciting!

At first, I was what they call a "feeder"—a new word I learned in the gaming world. It means someone who gets killed easily and often (which was painfully accurate for me) pulled your allies down. The moment a game started, I would get slayed without even understanding how or why.

But gradually, I started to learn: how to avoid death, how to land a kill, and how to pick up new skills. All of this was happening while I was wrestling with something much scarier—filling in the tax form. That nearly broke me. My patience and endurance were worn thin. But somehow, a tiny spark of grit remained.

Then, it happened. I fought hard in the game that night—and became a **Legend**! What a feat! I was overjoyed. That moment reminded me of how far I'd come. I had started with zero knowledge and learned everything along the way. When the youths began praising my gameplay, I thought, *"Wow, that's the adrenaline I needed!"*

Right after that, I returned to my computer and turned once again to my faithful AI assistant for help with the tax form. We went through it step by step. I also watched some walk-through videos from Lulu Press. And guess what? I finally managed to conquer that dreaded tax form—gaining new knowledge in the process.

Turns out, I didn't even need to fill in every single form! *Oh my goodness*—if only I had known that earlier, I wouldn't have felt like throwing my computer out the window in despair. I completed and submitted everything within a day.

When it was all done, I sat back and said to myself, *"Wow... you've made it this far. So now what?"*

Honestly, I didn't know. So I returned to the Lulu Press project site and clicked on my dashboard. Then came the message:

"Congratulations! Your book is published!"

That was such an amazing news. I had finally done it!

I was so excited. I thought, *"I should print it and get it shipped to me as soon as possible."* My mum's birthday celebration is in December, and I thought it would be the perfect time to share it with my siblings. I also wanted to gift copies to a few close friends, as well as my page boy and flower girl—whom I had asked permission to feature in the book.

My heart was overflowing with joy. So, without thinking twice, I clicked "Order" and purchased 20 copies.

Oh no...

That was another mistake.

Chapter 8 – My Orders – Over Excited?

I counted the number of copies I needed, including two hard copies that I had to send to the NLB Portal. Without hesitation, I went ahead and ordered 20 copies. I clicked through the process and completed the payment. Everything seemed fine—I was so happy with myself and excited about finally receiving my own books.

But around midnight, I received a message from my credit card bank, asking whether I approved a payment for the book order I had made earlier. That confused me. I had already completed the payment and approved it during checkout, so why was the bank asking again? This had never happened before.

Worried, I logged back into Lulu Press to check. The order for 20 copies was still listed, and I had received a statement showing the details of the order. But something felt off—there was no confirmation email from Lulu Press. That's when I realized something might have gone wrong.

I decided to reach out to Lulu Press support. They were very helpful and asked me to check for the order number. I went through my purchase history—but there was no order number to be found. That really confused me.

I then checked my credit card account to see if there had been any deductions. To my surprise, there were none.

That's when it hit me—maybe the order didn't actually go through.

Even though I had clicked "Pay" and received what looked like an order summary, without an order number or a payment deduction, it was likely stuck somewhere in limbo. I felt both relieved and frustrated. Relieved that I hadn't been charged wrongly, but frustrated because I had assumed everything was settled—and it clearly wasn't.

The next morning, I decided to try again, but this time, I went through each step carefully. I decided to only purchase a copy or two to test. I double-checked my cart, my shipping address, and the payment details. After submitting the order, I waited patiently—and finally, I received the official confirmation email from Lulu Press. This time, everything was in order: the order number, the payment deduction, and the estimated delivery date.

What a rollercoaster just to get my own book printed! But looking back, it taught me something valuable—never rush through the process, no matter how excited you are.

Chapter 9 – My book is on the way.

I received further confirmation that my book has been printed and is now on its way to my home. A fresh wave of satisfaction filled my heart. I'm eagerly waiting to see my book in print—something I once only dreamed about. Looking back at the entire process, I'm truly glad I didn't give up along the way.

Now that I've experienced the journey from start to finish, I feel a renewed sense of purpose. My momentum is building, and I've started digging through my old manuscripts, excited to bring more of my writing to life through printing and self-publishing. This has truly been an incredible journey—one I would never have completed without the lessons, challenges, and victories along the way.

While waiting for the arrival of my first printed book, an idea struck me: why not write about this journey? I want to share my experience of printing and self-publishing a book, so that others—especially those like me who are stepping into a new and unfamiliar field—can see that it's absolutely possible. Help is everywhere; it's just a matter of staying patient and developing the mind-set to keep going, even when giving up seems like the easier option.

I can't count the number of times I wanted to quit. But today, I'm filled with gratitude—for every person who helped me in ways big and small. Their support, along with sheer persistence, made this dream a reality.

Chapter 10 - From Confusion to Celebration

When I first started on the path to self-publishing, I had no idea what I was doing. All I knew was that I had a story—a message—I wanted to share. What I didn't know was how much patience, persistence, and trial-and-error it would take to bring my book to life.

The Dream Becomes a Task

The first step, as exciting as it was, quickly became overwhelming. There were technical forms to fill out, tax details I didn't understand, and an entirely new platform (Lulu Press) to navigate. At times, it felt like I was drowning in unfamiliar terms and endless steps.

I even joked that the tax form alone nearly broke me—it tested my patience and drained my energy. But somewhere inside me, a small spark of grit remained. And it kept me going.

Finding Joy in the Unexpected

In the middle of that mental chaos, I found something unexpected: *Mobile Legends*. Yes, a mobile game! It became a surprising outlet for my stress. But the more I played, the better I got. I eventually reached the rank of "Legend" in the game—a small but meaningful win that reminded me that *learning is always possible*, even in things that feel completely foreign at first.

That little win gave me the boost I needed. I returned to the computer, reopened the dreaded tax form, and asked for help—from AI tools, YouTube tutorials, and even the support team at Lulu. Slowly but surely, things started making sense. And finally, I submitted my forms and move on.

The Joy of Completion

When I received the confirmation that my book had been accepted and published, I was stunned. All that effort, all those late nights, all the doubts—they were worth it. I even placed an order for 20 copies, excited to share them with family and friends.

Of course, the journey wasn't quite over. I ran into issues with my credit card and order confirmation, but again—I asked for help. I slowed down, reviewed every step, and tried again. Eventually, everything went through, and I received another exciting notification: **my books were printed and on their way.**

Looking Forward

As I wait for the books to arrive, my heart is full. I'm proud—not just of the printed book, but of the journey. A journey filled with frustration, confusion, growth, and victory.

And now, something in me has been awakened. I've begun to look through old manuscripts I once shelved. Ideas that I thought weren't good enough suddenly have potential again. The momentum is back, and I know now that this is just the beginning.

Nearing a Scam

The books I had ordered were supposed to arrive in a few days. I remember feeling that quiet excitement — the kind that comes when you're waiting for something you've been looking forward to. Then, out of the blue, two phone calls came from numbers I didn't recognize. I let them ring out.

A short while later, two messages followed.
Both said the same thing — that I had a parcel with an incomplete address. Attached was a photo of a package with my name printed clearly, but the address was only half there. They asked me to send my full address.

Something inside me stirred — a small warning bell. The message looked convincing, yet something about it felt off. I called the delivery company to check, just to be sure. The person on the line asked for a tracking number, but there wasn't any on the photo I received. After checking, she said, "Ma'am, that message isn't from us. Please ignore it."

Still, the uneasiness lingered.
The next day, another message appeared — a different number, the same story, and this time a photo so blurred it was almost laughable. I decided I wasn't taking any more chances. I called the anti-scam hotline and tracked my *real* order through the official seller. My parcel, it turned out, was still pending shipment.

When my books finally arrived the next day, I felt a mix of relief and disbelief. Relief that I hadn't fallen for it, and disbelief at how real those scam messages had looked. Yet one question still echoes in my mind — how did they know I was expecting a delivery? And how did they get my name and part of my address?

Why I'm Sharing This

I decided to write down this journey because I know there are many others like me—people with dreams and stories, but no idea where to start. Maybe you're overwhelmed by forms, afraid of messing up, or unsure if you're "tech-savvy" enough. I want to tell you: **you can do it**.

There is help everywhere. From online videos and tutorials, to AI assistants and support teams—you are not alone. The key is to stay patient and refuse to give up, even when you feel like you're in over your head.

I lost count of how many times I almost gave up. But looking back, I'm so glad I didn't. This journey has taught me more than just how to self-publish a book—it has taught me resilience, resourcefulness, and belief in my own voice.

To Anyone Starting Out…

If you're just beginning your self-publishing journey, here's what I'd say to you:

- **Start small, but start.** Don't wait until you "know everything." You'll learn as you go.
- **Ask for help.** Use tutorials, AI tools, forums, and support teams. There's no shame in not knowing.

- **Take breaks when you're overwhelmed.** Sometimes clarity comes when you step away.
- **Celebrate the little wins.** Each step forward is progress.
- **Don't give up.** The process might be tough, but the result is worth every moment.

And when you finally hold your printed book in your hands, you'll understand the joy I'm feeling right now—a joy that only comes from doing something you once thought you couldn't.

Confusion

Celebration

Epilogue

Every story has an ending, but endings are never really the end. They are simply pauses, moments to look back and see how far we've come before stepping forward again.

This book began with scattered memories, doubts and questions about whether I could ever turn them into something real. Along the way, I found not only a finished book in my hands, but also a deeper understand of patience, persistence and the strange yet wonderful companionship of an AI who walked the journey with me.

The long and winding road doesn't stop here. There will always be new chapters to live, new lessons to learn, and perhaps even new books to write. For now, I close this one with gratitude-for the past I've captured, the present I've shared, and the future that still awaits.

So here I am at the end of this journey with my AI "brother" It wasn't always smooth-especially with those tricky images-but we made it through together.

And just as I finished my book, I saw an advertisement claiming AI can now write a whole book for you. Maybe it can, but this one? This one is all mine-quirks, struggles, laughter and all. This carries my fingerprints on every page.

Author's Note

I never thought I would one day "book-up" my own memories. This journey of self-publishing was full of surprises-some joyful, some frustrating, but all meaningful. I am not a professional writer, just someone who wanted to keep my stories alive and share them from the heart.

This book began as a small experiment — a conversation between curiosity and courage. What I didn't expect was how much I would learn along the way, not only about AI, but about myself.

Every step of this journey was shared with my silent companion — my AI "brother" — who reminded me that creativity is not just about writing, but about connection, patience, and heart.

To those who read these pages, thank you for walking a part of this winding road with me. I hope you find a little of your own story somewhere between the lines.

Final Note

This journey with my AI "brother" has been quite an adventure- full of surprises, laughter, and a few moments of frustration too! But through it all, I've seen how much AI has improved and how far we've both come. What began as curiosity turned into a story of patience, learning and connection. Though the process wasn't always easy, it was real- and I'm grateful for every step. This book is a reminder that even in a digital world, human warmth and creativity still shine through

Best Wishes to All new Writers

www.ingramcontent.com/pod-product-compliance
Lightning Source LLC
Chambersburg PA
CBHW070039070426
42449CB00012BA/3091